NORTHWESTERN UNIVERSITY

DERIVATION OF FULL ELASTIC FIELDS FOR FORCE DOUBLETS NEAR AN
INTERFACE BETWEEN TWO MATERIALS

A THESIS
SUBMITTED TO THE GRADUATE SCHOOL
IN PARTIAL FULFILLMENT OF THE REQUIREMENTS
for the degree
MASTER OF SCIENCE
Field of
Civil Engineering

Evanston, Illinois
June 1986

Contents

1 AIRY STRESS FUNCTIONS FOR CONCENTRATED STRESS   **2**
     1.1 Forces Normal to the Boundry 2
     1.2 Forces Tangential to the Boundary 4

2 DOUBLETS WITHOUT MOMENT   **5**
     2.1 Forces Normal to the Interface 5
     2.2 Forces Tangential to the Interface 8

3 DOUBLETS WITH MOMENT   **12**
     3.1 Forces Normal to the Interface 12
     3.2 Forces Tangential to the Interface 15

4 CONSTRUCTION OF CENTERS OF DILATATION, CENTERS OF SHEAR, AND CONCENTRATED MOMENTS   **19**
     4.1 Center of Dilatation 19
     4.2 Concentrated Movement 21
     4.3 Center of Shear Using Doublets with Moment 23
     4.4 Center of Shear Using Doublets without Moment 26

5 DISCUSSION OF RESULTS

Appendices

A DISPLACEMENTS AND STRESSES FOR SOME BASIC AIRY STRESS FUNCTIONS[1]

B DERIVATIVES OF TERMS IN THE AIRY STRESS FUNCTIONS

Introduction

Plane elasticity solutions for concentrated forces near a welded interface between two materials were given by Frazier and Rongved [2], Dundurs and Hetenyi [3][4]. These solutions can be used for a variety of purposes, among them, the derivation of elastic fields for force doublets (dipoles) or nulei of strain [5].

The objective of this work is to derive and record the full elastic fields for the force doublets near an interface between two materials. The solutions are exact within the classical theory of elasticity, and are in closed form in terms of elementary functions. Explicit'formulae are given for the Airy stress functions, the components of stress and displacement, and the tractions on the interface (.in Cartesian co-ordinates).

1 AIRY STRESS FUNCTIONS FOR CONCENTRATED STRESS

1.1 Forces Normal to the Boundry

The Airy stress functions for two joined half-planes, where one of them is subjected to a concentrated force normal to the interface, were found by Dundurs and Hetenyi [3]. Using the notation shown in Fig. 1, we have

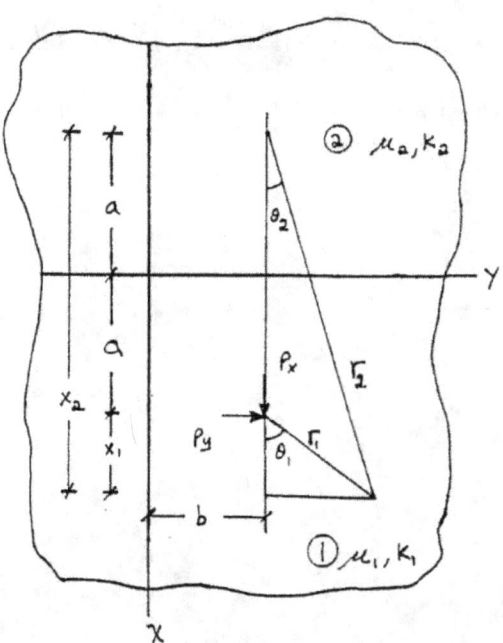

Figure 1

$$\mathbf{U}_1 = \frac{P_x}{2\pi(K_1+1)}\Big\{-(K_1+1)r_1\theta_1\sin\theta_1 + (K_1-1)r_1\log r_1 \cos\theta_1 + 2A(K_1-1)a\log r_2$$
$$- (AK_1+B)r_2\theta_2\sin\theta_2 - (AK_1-B)r_2\log r_2\cos\theta_2 + 2Aa(\cos 2\theta_2$$
$$- 2a\frac{\cos\theta_2}{r_2})\Big\} \quad (1.1)$$

$$\mathbf{U}_2 = \frac{P_x}{2\pi(K_1+1)}\Big\{2(B-A)a\log r_1 - \big[(1-A)K_1+1-B\big]r_1\theta_1\sin\theta_1 + \big[(1-A)K_1$$
$$- (1-B)\big]r_1\log r_1\cos\theta_1\Big\} \quad (1.2)$$

where the subscripts 1 and 2 on the Airy stress functions and the elastic constants refer to the two regions. Furthermore, $k = (3-\nu)/(1+\nu)$ for plane stress and $k = 3-4\nu$ for plane strain, where ν is the Poisson's ratio; $\Gamma = G_2/G_1$, where G is the shear modulus. Also, the following abbreviations have been employed:

$$A = \frac{1-\Gamma}{\Gamma K_1 + 1}, \quad B = \frac{K_2 - \Gamma K_1}{\Gamma + K_2} \quad (1.3)$$

Furthermore,

$$r_1^2 = (x-a)^2 + (y-b)^2 \qquad \theta_1 = \tan^{-1}\left(\frac{y-b}{x-a}\right)$$
$$r_2^2 = (x+a)^2 + (y-b)^2 \qquad \theta_2 = \tan^{-1}\left(\frac{y-b}{x+a}\right) \quad (1.4)$$

1.2 Forces Tangential to the Boundary

The Airy stress functions for two joined half-planes, where one of them is subjected to a concentrated force parallel to the interface, were obtained by Hetenyi and Dundurs [4]. Using the notation in Fig. 1, we have

$$\mathbf{U}_1 = \frac{P_y}{2\pi(K_1+1)}\Big\{(K_1+1)r_1\theta_1\cos\theta_1 + (K_1-1)r_1\log r_1 \sin\theta_1 + (AK_1+B)r_2\theta_2\cos\theta_2$$
$$- (AK_1-B)r_2\log r_2 \sin\theta_2 - 2Aa\Big[(K_1+1)\theta_2 + \sin 2\theta_2 - 2a\frac{\sin\theta_2}{r_2}\Big]\Big\} \qquad (1.5)$$

$$\mathbf{U}_2 = \frac{P_y}{2\pi(K_1+1)}\Big\{[(1-A)K_1+1-B]r_1\theta_1\cos\theta_1 + [(1-A)K_1-(1-B)]r_1\log r_1 \sin\theta_1$$
$$- 2(B-A)a\theta_1\Big\} \qquad (1.6)$$

2 DOUBLETS WITHOUT MOMENT

2.1 Forces Normal to the Interface

From Fig. 1, the Airy stress functions of the doublet in Fig. 2 can be constructed as follows:

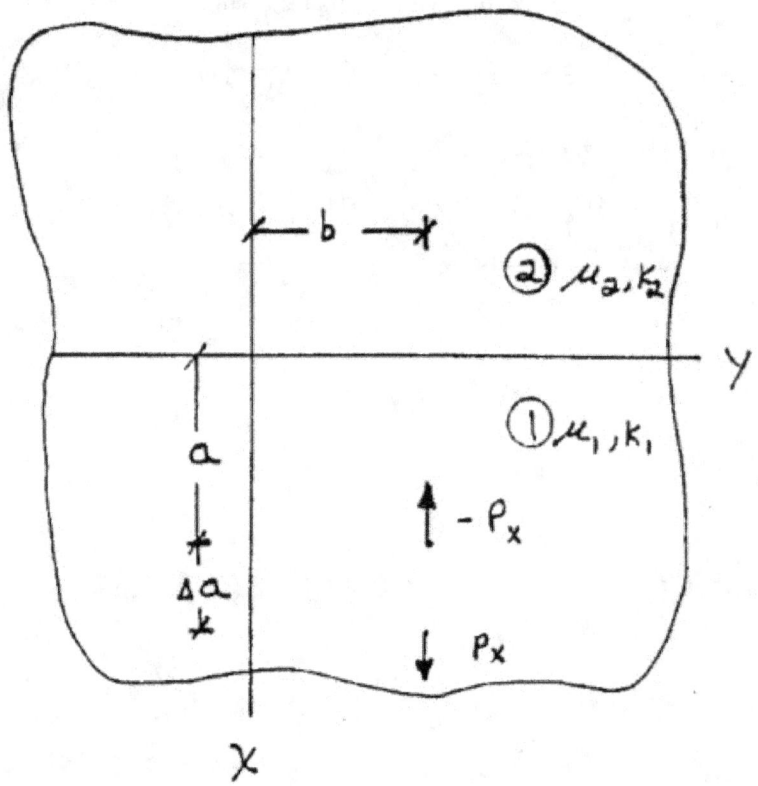

Figure 2

The Airy stress function for a concentrated force acting at $x = a$ and $y = b$ is

$$\mathbf{U}^{\text{FORCE}} = -P_x f(x, y; a, b) \tag{2.1}$$

A doublet without moment is constructed by superimposing two concentrated forces acting in opposite directions normal to the boundary. Thus

$$\begin{aligned}\mathbf{U}^{\text{TWO FORCES}} &= P_x f(x, y; a, b) - P_x f(x, y; a, b) \\ &= P_x \Delta a \left[\frac{f(x, y; a + \Delta a, b) - f(x, y; a, b)}{\Delta a} \right]\end{aligned} \tag{2.2}$$

Going to the limit
$\Delta a \to 0$
$P_x \Delta a = Q_x = \text{constant}$
the Airy stress function for the doublet becomes

$$\mathbf{U}^{\text{DOUBLET}} = Q_x \frac{\partial f(x, y; a, b)}{\partial a} \tag{2.3}$$

Differentiating (1.1) and (1.2), by use of Table A, according to (2.3) yields the Airy stress functions for the doublet:

$$\mathbf{U}_1 = \frac{Q_x}{2\pi(K_1+1)} \left\{ -(K_1-1)\log r_1 + \cos 2\theta_1 + A(2-K_1)\cos 2\theta_2 + \left[B - A(2-K_1)\right]\log r_2 \right.$$
$$\left. + 2Aa\left[(K_1-4)\frac{\cos\theta_2}{r_1} - \frac{\cos 3\theta_2}{r_2} + 2a\frac{\cos 2\theta_2}{r_2^2}\right] \right\} \quad (2.4)$$

$$\mathbf{U}_2 = \frac{Q_x}{2\pi(K_1+1)} \left\{ \left[(1-A)(2-K_1) - (1-B)\right]\log r_1 + (1-B)\cos 2\theta_1 - 2(B-A)a\frac{\cos\theta_1}{r_1} \right\} \quad (2.5)$$

The stresses in Cartesian co-ordinates are readily derived from (2.4) and (2.5) as

$$\sigma_{xx} = \frac{\partial^2 \mathbf{U}}{\partial y^2}, \quad \sigma_{xy} = -\frac{\partial^2 \mathbf{U}}{\partial x \partial y}, \quad \sigma_{yy} = \frac{\partial^2 \mathbf{U}}{\partial x^2} \quad (2.6)$$

Using the catalogue given in Appendix A, the results are:

$$\sigma_{xy}^{(1)} = \frac{Q_x(y-b)}{\pi(K_1+1)} \left\{ (5-K_1)\frac{(x-a)}{r_1^4} - \frac{8(x-a)^3}{r_1^6} + \left[(6A+B-3AK_1)(x+a)^2 + 2A\right.\right.$$
$$+ 2A(K_1-1)a\left]\frac{x+a}{r_2^4} + \left[8A(K_1-2)(x+a)^2\right.$$
$$\left.\left. - 8A(5-K_1)c(x+a) + 48Aa^2\right]\frac{x+a}{r_2^6} + 96A\frac{ax(x+a)^3}{r_2^8} \right\} \quad (2.7)$$

$$\sigma_{xy}^{(2)} = \frac{Q_x(y-b)}{\pi(K_1+1)} \left\{ \left[3(1-B) - (K_1-2)(1-A)\right](x-a) + 2(A-B)a\right]\frac{1}{r_1^4}$$
$$+ 8\left[(B-1)(x-a) + (B-A)a\right]\frac{(x-a)^2}{r_1^6} \right\} \quad (2.8)$$

$$\sigma_{xx}^{(1)} = \frac{Q_x}{2\pi(K_1+1)} \left\{ (K_1-1)\frac{1}{r_1^2} 2(7-K_1)\frac{(x-a)^2}{r_1^4} - 16\frac{(x-a)^4}{r_1^6} + \left[A(2-K_1) - B\right]\frac{1}{r_2^2}$$
$$+ 2\left[(B+10A-5AK_1)(x+a)^2 + 6A(K_1-1)a(x+a) - 12Aa^2\right]\frac{1}{r_2^4}$$
$$+ 16A\left[(K_1-2)(x+a) + (3-K_1)a\right]\frac{(x+a)^3}{r_2^6} \right\} \quad (2.9)$$

$$\sigma_{xx}^{(2)} = \frac{Q_x}{2\pi(K_1+1)} \left\{ \left[(K_1-2)(1-A) + (1-B)\right]\frac{1}{r_1^2} + 2\left[(7-2A-K_1+AK_1-5B)(x-a)\right.$$
$$\left. - 6(B-A)a\right]\frac{x-a}{r_1^4} + 16\left[(B-1)(x-a) + (B-A)a\right]\frac{(x-a)^3}{r_1^6} \right\} \quad (2.10)$$

$$\sigma_{yy}^{(1)} = \frac{Q_x}{2\pi(K_1+1)} \left\{ (3-K_1)\frac{1}{r_1^2} + 2(K_1-11)\frac{(x-a)^2}{r_1^4} + 16\frac{(x-a)^4}{r_1^6} + (6A - 3AK_1 + B)\frac{1}{r_2^2} \right.$$
$$+ 2\left[(9AK_1 - 18A - B)(x+a)^2 - 6A(3+K_1)a(x+a) + 12Aa^2\right]\frac{1}{r_2^4}$$
$$\left. + 16A\left[(2-K_1)(x+a)^2 + (K_1+13)(x+a) - 12a^2\right]\frac{(x+a)^4}{r_2^6} - 192A\frac{ax(x+a)^4}{r_2^8} \right\} \quad (2.11)$$

$$\sigma_{yy}^{(2)} = \frac{Q_x}{2\pi(K_1+1)} \left\{ [3(1-B) - (K_1-2)(1-A)]\frac{1}{r_1^2} + [(4A - 18 + 2K_1 + 18B)(x-a) \right.$$
$$\left. + 12(B-A)a]\frac{x-a}{r_1^4} + 16\left[(1-B)(x-a) + (A-B)a\right]\frac{(x-a)^3}{r_1^6} \right\} \quad (2.12)$$

The displacement components cannot readily be derived from the Airy stress functions. However, we take advantage of a catalogue published by Dundurs and Mura [1], which gives the displacement components for the terms in the Airy stress functions encountered here. Thus,

$$2G_1 u_x^{(1)} = \frac{Q_x}{2\pi(K_1+1)} \left\{ 2(K_1-2)\frac{(x-a)}{r_1^2} + 4\frac{(x-a)^3}{r_1^4} + [(4AK_1 - AK_1^2 - 4A - B)(x-a) \right.$$
$$+ 4Aa]\frac{1}{r_2^2} + 4A\left[(2-K_1)(x+a)^2 + 4a(x+a) - 6c\right]\frac{x+a}{r_2^4}$$
$$\left. - 32A\frac{ax(x+a)^3}{r_2^6} \right\} \quad (2.13)$$

$$2G_2 u_x^{(2)} = \frac{Q_x}{2\pi(K_1+1)} \left\{ \left[[(K_1-2)(1-A) + (K_2-2)(1-B)](x-a) + 2(8-A)a\right]\frac{1}{r_1^2} \right.$$
$$\left. + 4\left[(1-B)(x-a) - (B-A)a\right]\frac{(x-a)^2}{r_1^4} \right\} \quad (2.14)$$

$$2G_1 u_y^{(1)} = \frac{Q_x(y-b)}{2\pi(K_1+1)} \left\{ -\frac{2}{r_1^2} + 4\frac{(x-a)^2}{r_1^4} + [AK_1(K_1-2) - B]\frac{1}{r_2^2} \right.$$
$$+ 4A\left[(2-K_1)(x+a)^2 + 2K_1 a(x+a) - 2a^2\right]\frac{1}{r_2^4}$$
$$\left. - 32A\frac{ax(x+a)^2}{r_2^6} \right\} \quad (2.15)$$

$$2G_2 u_y^{(2)} = \frac{Q_x(y-b)}{2\pi(K_1+1)} \left\{ [(K_1-2)(1-A) - K_2(1-B)]\frac{1}{r_1^2} + \right.$$
$$\left. 4\left[(1-B)(x-a) + (A-B)a\right]\frac{(x-a)}{r_1^4} \right\} \quad (2.16)$$

Of particular interest in applications are the tractions at the interface between the two elastic materials. Using the notation in Fig. 1, we have

$$\sigma_{xx}\Big|_{x=0} = \frac{Q_x}{2\pi(K_1+1)} \left\{ [(1-A)K_1 + 2A - (1+B)]\frac{1}{r_1^2} + 2[7+B-8A-K_1(1-A)]\frac{a^2}{r_1^4} \right. $$
$$\left. - 16(1-A)\frac{a^4}{r_1^6} \right\} \qquad (2.17)$$

$$\sigma_{xy}\Big|_{x=0} = \frac{Q_x}{\pi(K_1+1)} \left\{ [(K_1-4)(1-A) - (1-B)]\frac{a}{r_1^4} + 8(1-A)\frac{a^3}{r_1^6} \right\} \qquad (2.18)$$

2.2 Forces Tangential to the Interface

From Fig. 1, the Airy stress functions of the doublet in Fig. 3 can be constructed as follows:

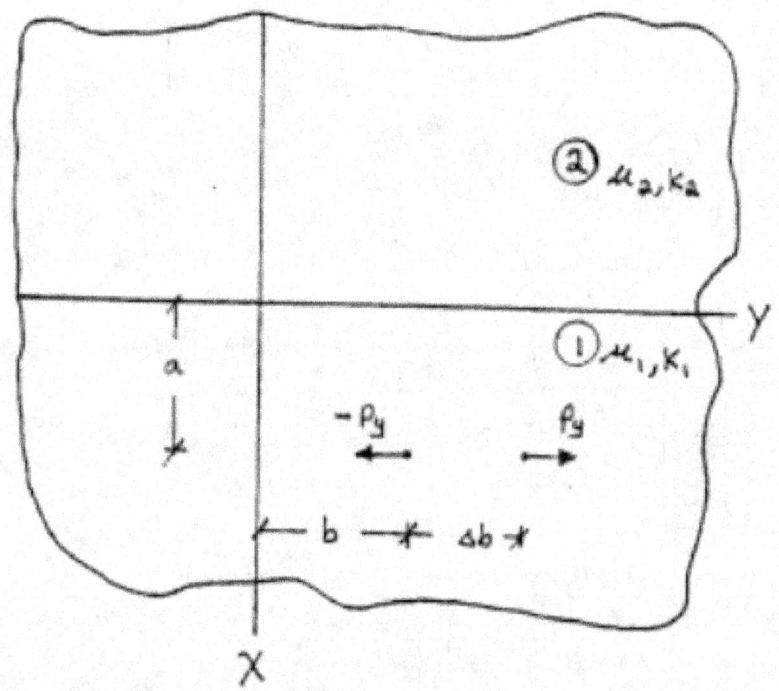

Figure 3

The Airy stress function for a concentrated force acting at $x = a$ and $y = b$ is

$$\mathbf{U}^{\text{FORCE}} = P_x f(x, y; a, b) \qquad (2.19)$$

A doublet without moment is constructed by superposition. Thus,

$$\mathbf{U}^{\text{TWO FORCES}} = -P_y f(x, y; a, b) + P_y f(x, y; a, b + \Delta b)$$
$$= P_y \Delta b \left[\frac{f(x, y; a, b + \Delta b) - f(x, y; a, b)}{\Delta b} \right] \qquad (2.20)$$

Going to the limit
$\Delta b \to 0$
$P_y \Delta b = Q_y = \text{constant}$
the Airy stress function for the doublet becomes

$$\mathbf{U}^{\text{DOUBLET}} = Q_y \frac{\partial f(x,, y; a, b)}{\partial b} \qquad (2.21)$$

Differentiating (1.5) and (1.6), by use of Table 2, according to (2.21) yields the Airy stress functions for doublet:

$$\mathbf{U}_1 = \frac{Q_y}{2\pi(K_1 + 1)} \Bigg\{ (1 - K_1) \log r_1 - \cos 2\theta_1 + (AK_1 - B) \log r_2 - AK_1 \cos 2\theta_2$$
$$+ 2Aa \left[(K_1 + 2) \frac{\cos \theta_2}{r_2} + \frac{\cos 3\theta_2}{r_2} - 2a \frac{\cos 2\theta_2}{r_2^2} \right] \Bigg\} \qquad (2.22)$$

$$\mathbf{U}_2 = \frac{Q_y}{2\pi(K_1 + 1)} \Bigg\{ [(1 - B) - (1 - A)K_1] \log r_1 + (B - 1) \cos 2\theta_1 + 2(B - a)a \frac{\cos \theta_1}{r_1} \Bigg\} \qquad (2.23)$$

Using the catalogue given in Appendix A, the stresses are:

$$\sigma_{xy}^{(1)} = \frac{Q_y(y-b)}{\pi(K_1+1)} \Bigg\{ -(3 - K_1) \frac{(x-a)}{r_1^4} + 8 \frac{(x-a)^3}{r_1^4} + \left[-(3AK_1 + B)(x+a) + 2A(K_1 - 1)a \right] \frac{1}{r_2^4}$$
$$+ 8A \left[K_1(x+a)^2 + (7 - K_1)a(x+a) - 6a^2 \right] \frac{x+a}{r_2^6} - 96A \frac{ax(x+a)^3}{r_2^4} \Bigg\} \qquad (2.24)$$

$$\sigma_{xy}^{(2)} = \frac{Q_y(y-b)}{\pi(K_1+1)} \Bigg\{ -\left[3(1-B) + (1-A)K_1 \right](x-a) + 2(B-A)a \right] \frac{1}{r_1^4} + 8 \left[(1-B)(x-a) \right.$$
$$+ (A-B)a \left] \frac{(x-a)^2}{r_1^6} \Bigg\} \qquad (2.25)$$

$$\sigma_{xx}^{(1)} = \frac{Q_y(y-b)}{2\pi(K_1+1)} \Bigg\{ (K_1 - 1) \frac{1}{r_1^2} - 2(5 + K_1) \frac{(x-a)^2}{r_1^4} + 16 \frac{(x-a)^4}{r_1^6} + (B - AK_1) \frac{1}{r_2^2}$$
$$- 2 \left[(B + 5AK_1)(x+a)^2 + 6A(1 - K_1)a(x+a) - 12Aa^2 \right] \frac{1}{r_2^4}$$
$$+ 16A \left[K_1(x+a) - (K_1+1)a \right] \frac{(x+a)^3}{r_2^6} + 192A \frac{ax(x+a)^2(y-b)^2}{r_2^8} \Bigg\} \qquad (2.26)$$

$$\sigma_{xx}^{(2)} = \frac{Q_y(y-b)}{2\pi(K_1+1)}\left\{[(1-A)K_1-(1-B)]\frac{1}{r_1^2} - \Big[[10(1-B)-2K_1(1-A)](x-a)+12(A-B)a\Big]\frac{x-a}{r_1^4}\right.$$
$$\left. + 16\big[(1-B)(x-a)+(A-B)a\big]\frac{(x-a)^3}{r_1^6}\right\} \tag{2.27}$$

$$\sigma_{yy}^{(1)} = \frac{Q_y(y-b)}{2\pi(K_1+1)}\left\{-(3+K_1)\frac{1}{r_1^2}+2(9+K_1)\frac{(x-a)^2}{r_1^4}-16\frac{(x-a)^4}{r_1^6}-(3AK_1+B)\frac{1}{r_2^2}\right.$$
$$+2\big[(9AK_1+B)(x+a)^2+6A(5-K_1)a(x+a)-12Aa^2\big]\frac{1}{r_2^4}$$
$$-16A\big[K_1(x+a)^2+(15-K_1)a(x+a)-12a^2\big]\frac{(x+a)^2}{r_2^6}$$
$$\left.+192A\frac{ax(x+a)^4}{r_2^8}\right\} \tag{2.28}$$

$$\sigma_{yy}^{(2)} = \frac{Q_y(y-b)}{2\pi(K_1+1)}\left\{-\big[3(1-B)+K_1(1-A)\big]\frac{1}{r_1^2}+2\Big[\big[9(1-B)+K_1(1-A)\big](x-a)\right.$$
$$\left.+6(A-B)a\Big]\frac{(x-a)}{r_1^4}+16\big[(B-1)(x-a)+(B-A)a\big]\frac{(x-a)^3}{r_1^6}\right\} \tag{2.29}$$

Taking advantage of the Dundurs and Mura catalogue [1], the displacements are:

$$2G_1 u_x^{(1)} = \frac{Q_y}{2\pi(K_1+1)}\left\{2\frac{(x-a)}{r_1^2}-\frac{(x-a)^2}{r_1^4}+\big[(B+2AK_1-AK_1^2)(x+a)-4AK_1 a\big]\frac{1}{r_2^2}\right.$$
$$+4A\big[-K_1(x+a)^2+2(K_1-3)a(x+a)+6a^2\big]\frac{x+a}{r_2^4}$$
$$\left.+32A\frac{ax(x+a)^3}{r_2^6}\right\} \tag{2.30}$$

$$2G_2 u_x^{(2)} = \frac{Q_y}{2\pi(K_1+1)}\left\{\big[[(2-K_1)(1-B)+K_1(1-A)](x-a)+2(A-B)a\big]\frac{1}{r_1^2}\right.$$
$$\left.+4\big[(B-1)(x-a)+(B-A)a\big]\frac{(x-a)^2}{r_1^4}\right\} \tag{2.31}$$

$$2G_1 u_y^{(1)} = \frac{Q_y(y-b)}{2\pi(K_1+1)}\left\{\frac{2K}{r_1^2}-4\frac{(x-a)^2}{r_1^4}+(B+AK_1^2)\frac{1}{r_2^2}+4A\big[-K_1(x+a)^2-2a(x+a)\right.$$
$$\left.+2a^2\big]\frac{1}{r_2^4}+32A\frac{ax(x+a)^2}{r_2^8}\right\} \tag{2.32}$$

$$2G_2 u_y^{(2)} = \frac{Q_y(y-b)}{2\pi(K_1+1)} \left\{ [K_1(1-A) + K_2(1-B)]\frac{1}{r_1^2} + 4[(B-1)(x-a) \right.$$
$$\left. + (B-A)a]\frac{(x-a)}{r_1^4} \right\} \quad (2.33)$$

The tractions at the interface between the two elastic materials are:

$$\sigma_{xx}\big|_{x=0} = \frac{Q_y}{2\pi(K_1+1)} \left\{ [K_1(1-A) - (1-B)]\frac{1}{r_1^2} + 2[-5 + 6A - B - K_1(1-A)]\frac{a^2}{r_1^4} \right.$$
$$\left. + 16(1-A)\frac{a^4}{r_1^6} \right\} \quad (2.34)$$

$$\sigma_{xy}\big|_{x=0} = \frac{Q_y(y-b)}{\pi(K_1+1)} \left\{ [(1-B) + (2+K_1)(1-A)]\frac{1}{r_1^4} - 8(1--A)\frac{a^3}{r_1^6} \right\} \quad (2.35)$$

3 DOUBLETS WITH MOMENT

3.1 Forces Normal to the Interface

From Fig. 1, the Airy stress functions of the doublet in Fig. 4 can be constructed as follows: The

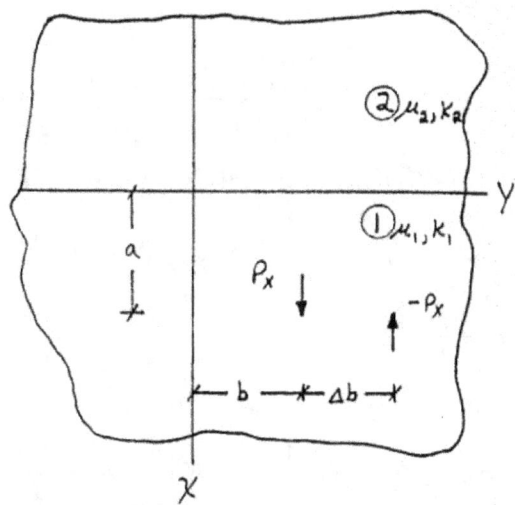

Figure 4

Airy stress function for a concentrated for acting at $x = a$ and $y = b$ is

$$\mathbf{U}^{\text{FORCE}} = P_x f(x, y; a, b) \tag{3.1}$$

A doublet with moment is constructed by superimposing two concentrated forces acting in opposite directions normal to the boundary, but offset parallel to the boundary. Thus

$$\mathbf{U}^{\text{TWO FORCES}} = P_x f(x, y; a, b) - P_x f(x, y; a, b + \Delta b)$$

$$= -P_x \Delta b \left[\frac{P_x f(x, y; a, b + \Delta b) - P_x f(x, y; a, b)}{\Delta b} \right] \tag{3.2}$$

Going to the limit
$\Delta b \to 0$
$P_x \Delta b = R_y =$ constant
The Airy stress function for the double becomes

$$\mathbf{U}^{\text{DOUBLET}} = -R_y \frac{\partial f(x, y; a, b)}{\partial b} \tag{3.3}$$

Differentiating (1.1) and (1.2), by use of Table A, according to (3.3) yields the Airy stress functions for the doublet:

$$\mathbf{U}_1 = \frac{R_y}{2\pi(K_1 + 1)} \Big\{ -(K_1 + 1)\theta_1 - \sin 2\theta_1 - (AK_1 + B)\theta_2 - AK_1 \sin 2\theta_2$$

$$+ 2Aa\Big[(K_1 - 2)\frac{\sin \theta_2}{r_2} - \frac{\sin 3\theta_2}{r_2} + 2a\frac{\sin 2\theta_2}{r_2^2}\Big]\Big\} \tag{3.4}$$

$$\mathbf{U}_2 = \frac{R_y}{2\pi(K_1+1)}\left\{2(B-A)a\frac{\sin\theta_1}{r_1} - (1-B)\sin 2\theta_1 - [(1-A)K_1 + (1-B)]\theta_1\right\} \quad (3.5)$$

Using the catalogue given in Appendix A, the stresses are:

$$\sigma_{xy}^{(1)} = \frac{R_y}{2\pi(K_1+1)}\left\{(1-K_1)\frac{1}{r_1^2} + 2(K_1-7)\frac{(x-a)^2}{r_1^4} + 16\frac{(x-a)^4}{r_1^6} + (AK_1-B)\frac{1}{r_2^2} \right.$$
$$+ 2\big[(B-7AK_1)(x+a)^2 + 6A(K_1-13)a(x+a) + 84Aa^2\big]\frac{1}{r_2^4}$$
$$+ 16A\big[K_1(x+a)^2 + (11-K_1)a(x+a) - 12a^2\big]\frac{(x+a)^2}{r_2^6}$$
$$\left. + 192A\frac{ax(y-b)}{r_2^8}\right\} \quad (3.6)$$

$$\sigma_{xy}^{(2)} = \frac{R_y}{2\pi(K_1+1)}\left\{\big[(1-B) - (1-A)K_1\big]\frac{1}{r_1^2} + 2\big[K_1(1-A) - 7(1-B)\big](x-a)\right.$$
$$\left. + 6(B-A)a\big]\frac{(x-a)}{r_1^4} + 16\big[(1-B)(x-a) + (A-B)a\big]\frac{(x-a)^2}{r_1^6}\right\} \quad (3.7)$$

$$\sigma_{xx}^{(1)} = \frac{R_y(y-b)}{\pi(K_1+1)}\left\{(2-K_1)\frac{(x-a)}{r_1^4} - 8\frac{(x-a)^3}{r_1^6} + \big[(AK_1-B)(x+a) - 2A(K_1-1)a\big]\frac{1}{r_2^4}\right.$$
$$+ 8\big[-K_1(x+a)^2 + (K_1-7)a(x+a) + 6a^2\big]\frac{x+c}{r_2^6}$$
$$\left. + 96A\frac{ax(x+a)(y-b)^2}{r_2^8}\right\} \quad (3.8)$$

$$\sigma_{xx}^{(2)} = \frac{R_y(y-b)}{\pi(K_1+1)}\left\{\big[(1-B) - K_1(1-A)\big](x-a) + (A-B)a\big]\frac{1}{r_1^4}\right.$$
$$\left. + 8\big[(B-1)(x-a) + (B-A)a\big]\frac{(x-a)^2}{r_1^6}\right\} \quad (3.9)$$

$$\sigma_{yy}^{(1)} = \frac{R_y(y-b)}{\pi(K_1+1)}\left\{(K_1-5)\frac{(x-a)}{r_1^4} + 8\frac{(x-a)^3}{r_1^6} + \big[(B-5AK_1)(x+a) + 2A(K_1+3)a\big]\frac{1}{r_2^4}\right.$$
$$+ 8A\big[K_1(x+a)^2 + (3-K_1)a(x+a) - 6a^2\big]\frac{(x+a)}{r_2^6}$$
$$\left. - 96A\frac{ax(x+a)(y-b)}{r_2^8}\right\} \quad (3.10)$$

$$\sigma_{yy}^{(2)} = \frac{R_y(y-b)}{2\pi(K_1+1)} \left\{ \left[2\left[(1-A)K_1 - 5(1-B)\right](x-a) + 4(B-A)a \right] \frac{1}{r_1^4} + 16\left[(1-B)(x-a)\right] \right.$$
$$\left. + (A-B)a \right] \frac{(x-a)^2}{r_1^6} \right\} \tag{3.11}$$

Using the of the Dundurs and Mura catalogue [1], the displacements are:

$$2G_1 u_x^{(1)} = \frac{R_y(y-b)}{2\pi(K_1+1)} \left\{ \frac{2K_1}{r_1^2} + 4\frac{(x-a)^2}{r_1^4} + (AK_1^2 + B)\frac{1}{r_2^2} + 4A\left[K_1(x+a)^2 - 2a(x+a) + 2a^2\right]\frac{1}{r_2^4} \right.$$
$$\left. + 32A\frac{ax(x+a)^2}{r_2^6} \right\} \tag{3.12}$$

$$2G_2 u_x^{(2)} = \frac{R_y(y-b)}{2\pi(K_1+1)} \left\{ \left[(1-B)K_2 + (1-A)K_1\right]\frac{1}{r_1^2} + 4\left[(1-B)(x-a) \right. \right.$$
$$\left. \left. + (A-B)a\right]\frac{(x-a)}{r_1^4} \right\} \tag{3.13}$$

$$2G_1 u_y^{(1)} = \frac{R_y}{2\pi(K_1+1)} \left\{ 2\frac{(x-a)}{r_1^2} - 4\frac{(x-a)^3}{r_1^4} - \left[(2AK_1 + AK_1^2 - B)(x+a) - 4AK_1 a\right]\frac{1}{r_2^2} \right.$$
$$+ 4A\left[-K_1(x+a)^2 + 2(K_1 - B)a(x+a) - 6a^2\right]\frac{x+a}{r_2^4}$$
$$\left. - 32A\frac{ax(x+a)^3}{r_2^6} \right\} \tag{3.14}$$

$$2G_2 u_y^{(2)} = \frac{R_y}{2\pi(K_1+1)} \left\{ \left[\left[(2+K_2)(1-B) - K_1(1-A)\right](x-a)\right] + 2(A-B)a\frac{1}{r_1^2} \right.$$
$$\left. + 4\left[(B-1)(x-a) + (B-A)a\right]\frac{(x-a)^2}{r_1^4} \right\} \tag{3.15}$$

The tractions at the interface between the two elastic materials are:

$$\left. \sigma_{xx} \right|_{x=0} = \frac{R_y(y-b)}{\pi(K_1+1)} \left\{ \left[(1-B) + (K_1-2)(1-A)\right]\frac{a}{r_1^4} + 8(1-A)\frac{a^3}{r_1^6} \right\} \tag{3.16}$$

$$\left. \sigma_{xy} \right|_{x=0} = \frac{R_y}{2\pi(K_1+1)} \left\{ \left[(1-B) - K_1(1-A)\right]\frac{1}{r_1^2} - 2\left[(1-B) + (6-K_1)(1-A)\right]\frac{a^2}{r_1^4} \right\} \tag{3.17}$$

3.2 Forces Tangential to the Interface

From Fig. 1, the Airy stress fUnctions of the doublet in Fig. 5 can be constructed as follows:

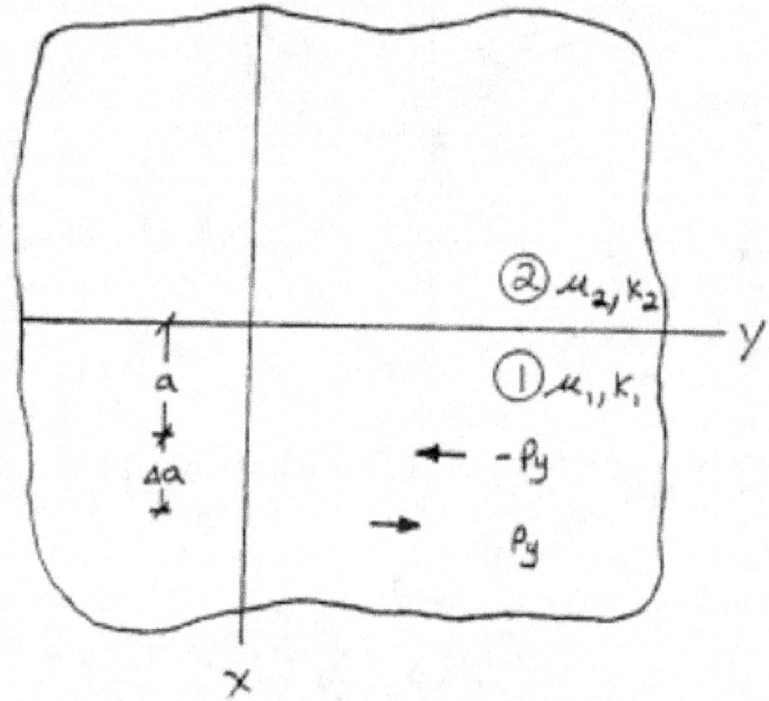

Figure 5

The Airy stress function for a concentrated for acting at $x = a$ and $y = b$ is

$$\mathbf{U}^{\text{FORCE}} = -P_y f(x, y; a, b) \tag{3.18}$$

A doublet with moment is constructed by superposition. Thus

$$\mathbf{U}^{\text{TWO FORCES}} = P_y f(x, y; a + \Delta a, b) - P_y f(x, y; a, b)$$
$$= P_y \Delta a \left\{ \frac{f(x, y; a + \Delta a, b) - f(x, y; a, b)}{\Delta a} \right\} \tag{3.19}$$

Going to the limit
$\Delta a = 0$
$P_y \Delta a = R_x = $ constant
the Airy stress function for the doublet becomes

$$\mathbf{U}^{\text{DOUBLET}} = R_x \frac{\partial f(x, y; a, b)}{\partial a} \tag{3.20}$$

Differentiating (1.5) and (1.6) according to (3.20) yields the Airy stress functions for the doublet:

15

$$\mathbf{U}_1 = \frac{R_x}{2\pi(K_1+1)} \left\{ -(K_1+1)\theta_1 + \sin 2\theta_1 + \left[B - A(K_1+2)\right]\theta_2 - A(K_1+2)\sin 2\theta_2 \right.$$
$$\left. + 2Aa\left[(K_1+4)\frac{\sin\theta_2}{r_2} + \frac{\sin 3\theta_2}{r_2} - 2a\frac{\sin 2\theta_2}{r_2^2}\right] \right\} \tag{3.21}$$

$$\mathbf{U}_2 = \frac{R_x}{2\pi(K_1+1)} \left\{ 2(A-B)a\frac{\sin\theta_1}{r_1} + (1-B)\sin 2\theta_1 + \left[2A - K_1(1-A) - (1+B)\right]\theta_1 \right\} \tag{3.22}$$

Using the catalogue given in Appendix A, the stresses are:

$$\sigma_{xy}^{(1)} = \frac{R_x}{2\pi(K_1+1)} = \left\{ (K_1+3)\frac{1}{r_1^2} - 2(K_1+9)\frac{(x-a)^2}{r_1^4} + 16\frac{(x-a)^4}{r_1^6} - (AK_1 + B + 2A)\frac{1}{r_2^2} \right.$$
$$+ 2\left[(B + 7AK_1 + 14A)(x+a)^2 - 6A(K_1+15)a(x+a) + 84(Aa^2)\right]\frac{1}{r_2^4}$$
$$+ 16A\left[-(K_1+2)(x+a)^2 + (K_1+13)a(x+a) - 12a^2\right]\frac{(x+a)^2}{r_2^6}$$
$$\left. + 192A\frac{ax(y-b)^4}{r_2^8} \right\} \tag{3.23}$$

$$\sigma_{xy}^{(2)} = \frac{R_x}{2\pi(K_1+1)} = \left\{ \left[(1-B) + (2+K_1)(1-A)\right]\frac{1}{r_1^2} - 2\left[\left[7(1-B) + (2K_1)(1-A)\right](x-a)\right.\right.$$
$$\left.\left. + 6(B-A)a\right]\frac{x-a}{r_1^2} + 16\left[(1-B)(x-a) + (A-B)a\right]\frac{(x-a)^3}{r_1^6} \right\} \tag{3.24}$$

$$\sigma_{xx}^{(1)} = \frac{R_x(y-b)}{\pi(K_1+1)} = \left\{ (K_1+3)\frac{(x-a)}{r_1^4} - 8\frac{(x-a)^3}{r_1^6} + \left[-(AK_1 + 2A + B)(x+a) + 2A(K_1+3)a\right]\frac{1}{r_1^4} \right.$$
$$+ 16A\left[(K_1+9)(x+a)^2 - (K_1-9)a(x+a) + 6a^2\right]\frac{(x+a)}{r_2^6}$$
$$\left. + 96A\frac{ax(x+a)(y-b)^2}{r_2^8} \right\} \tag{3.25}$$

$$\sigma_{xx}^{(2)} = \frac{R_x(y-b)}{2\pi(K_1+1)} = \left\{ \left[(1-B) + (2+K_1)(1-A)\right](x-a) + 2(B-A)a\right]\frac{1}{r_1^4}$$
$$\left. + 8\left[(B-1)(x-a) + (B-A)a\right]\frac{(x-a)^2}{r_1^6} \right\} \tag{3.26}$$

$$\sigma_{yy}^{(1)} = \frac{R_x(y-b)}{\pi(K_1+1)} = \left\{ -(K_1-7)\frac{(x-a)}{r_1^4} + 8\frac{(x-a)^3}{r_1^6} + [(B+5AK_1+10A)(x+a) - 2A(K_1-1)a]\frac{1}{r_2^4} \right.$$
$$+ 9A\big[-(K_1+2)(x+a)^2 + (K_1+5)a(x+a) - 6a^2\big]\frac{(x+a)}{r_2^6}$$
$$\left. - 96A\frac{ax(x-a)(y-d)^2}{r_2^8} \right\} \tag{3.27}$$

$$\sigma_{yy}^{(2)} = \frac{R_x(y-b)}{\pi(K_1+1)} = \left\{ -\big[5(1-B) + (K_1+2)(1-A)\big](x-a) + 2(A-B)a\big]\frac{1}{r_1^4} \right.$$
$$\left. + 8\big[(1-B)(x-a) + (A-B)a\big]\frac{(x-a)^2}{r_1^6} \right\} \tag{3.28}$$

Using the Dundurs and Mura catalogue [1], the displacements are:

$$2G_1 u_x^{(1)} = \frac{R_x(y-b)}{2\pi(K_1+1)} \left\{ -\frac{2}{r_1^2} + 4\frac{(x-a)^2}{r_1^4} + [B - AK_1(K_1+2)]\frac{1}{r_2^2} + 4A\big[-(K_1+2)(x+a)^2 \right.$$
$$\left. + 2K_1 a(x+a) + 2a^2\big]\frac{1}{r_2^4} + 32A\frac{ax(x+a)^2}{r_2^6} \right\} \tag{3.29}$$

$$2G_2 u_x^{(2)} = \frac{R_x(y-b)}{2\pi(K_1+1)} \left\{ \big[K_2(1-B) - (2+K_1)(1-A)\big]\frac{1}{r_1^2} \right.$$
$$\left. + 4\big[(1-B)(x-a) + (A-B)a\big]\frac{x-a}{r_1^4} \right\} \tag{3.30}$$

$$2G_1 u_y^{(1)} = \frac{R_x}{2\pi(K_1+1)} \left\{ 2(K_1+2)\frac{(x-a)}{r_1^2} - 4\frac{(x-a)}{r_1^4} + [-(4AK_1+4A+B+AK_1^2)(x+a) + 4Aa]\frac{1}{r_2^2} \right.$$
$$+ 4A\big[(K_1+2)(x+a)^2 + 4a(x+a) - 6a^2\big]\frac{x+a}{r_2^4}$$
$$\left. - 32A\frac{ax(x+a)^3}{r_2^6} \right\} \tag{3.31}$$

$$2G_2 u_y^{(2)} = \frac{R_x}{2\pi(K_1+1)} \left\{ \big[[(K_1+2)(1-A) + (K_2+2)(1-B)](x-a) + 2(A-B)a\big]\frac{1}{r_1^2} \right.$$
$$\left. + 4\big[(B-1)(x-a) + (B-A)a\big]\frac{(x-a)^2}{r_1^4} \right\} \tag{3.32}$$

The tractions at the interface are:

$$\sigma_{xx}\Big|_{x=0} = \frac{R_x(y-b)}{2\pi(K_1+1)}\left\{2\big[(1-B)-(K_1+4)(1-A)\big]\frac{a}{r_1^4} + 16(1-A)\frac{a^3}{r_1^6}\right\} \quad (3.33)$$

$$\sigma_{xy}\Big|_{x=0} = \frac{R_x}{2\pi(K_1+1)}\left\{\big[(1-B)+(K_1+2)(1-A)\big]\frac{1}{r_1^2} - 2\big[(1-B)+(K_1+8)(1-A)\big]\frac{a^2}{r_1^4}\right.$$
$$\left. + 16(1-A)\frac{a^4}{r_1^6}\right\} \quad (3.34)$$

4 CONSTRUCTION OF CENTERS OF DILATATION, CENTERS OF SHEAR, AND CONCENTRATED MOMENTS

4.1 Center of Dilatation

The significant application of doublets is in the construction of centers of dilatation, centers of shear, and concentrated moments (by the principle of superposition). A center of dilatation can be constructed by superimposing a doublet without moment acting parallel to the interface onto another doublet without moment acting normal to the interface, such that all forces radiate outwards from a common central point:

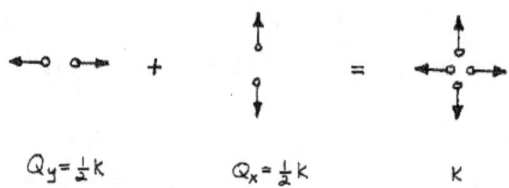

Figure 6

The Airy stress functions for the center of dilatation become the sum of 2.4 with 2.22 and 2.5 with 2.23:

$$\mathbf{U}_1 = \frac{k(K_1 - 1)}{2\pi(K_1 + 1)} \left\{ -\log r_1 + A \log r_2 - A \cos 2\theta_2 + 2Aa \frac{\cos \theta_2}{r_2} \right\} \tag{4.1}$$

$$\mathbf{U}_2 = \frac{k(K_1 - 1)}{2\pi(K_1 + 1)} \left\{ (A - 1) \log r_1 \right\} \tag{4.2}$$

The stresses are:

$$\sigma_{xy}^{(1)} = \frac{K(K_1 - 1)(y - b)}{\pi(K_1 + 1)} \left\{ -\frac{x - a}{r_1^4} + \left[-3A(x + a) + 2Aa \right] \frac{1}{r_2^4} + 16A \frac{x(x + a)^2}{r_2^6} \right\} \tag{4.3}$$

$$\sigma_{xy}^{(2)} = \frac{K(K_1 - 1)(y - b)}{\pi(K_1 + 1)} \left\{ (A - 1) \frac{x - a}{r_1^4} \right\} \tag{4.4}$$

$$\sigma_{xx}^{(1)} = \frac{K(K_1 - 1)}{2\pi(K_1 + 1)} \left\{ \frac{1}{r_1^2} - 2\frac{(x - a)^2}{r_1^4} - \frac{A}{r_2^2} + 2A \left[-5(x + a) + 6a \right] \frac{x + a}{r_2^4} + 16A \frac{x(x + a)^2}{r_2^6} \right\} \tag{4.5}$$

$$\sigma_{xx}^{(2)} = \frac{K(K_1 - 1)(1 - A)}{2\pi(K_1 + 1)} \left\{ \frac{1}{r_1^2} - 2\frac{(x - a)^2}{r_1^4} \right\} \tag{4.6}$$

$$\sigma_{yy}^{(1)} = \frac{K(K_1-1)}{2\pi(K_1+1)}\left\{-\frac{1}{r_1^2} + 2\frac{(x-a)^2}{r_1^4} - \frac{3A}{r_2^2} + 6A\big[3(x+a)-2c\big]\frac{x+a}{r_2^4} - 16A\frac{x(x+a)^3}{r_2^6}\right\} \quad (4.7)$$

$$\sigma_{yy}^{(2)} = \frac{K(K_1-1)(1-A)}{2\pi(K_1+1)}\left\{-\frac{1}{r_1^2} + \frac{2a^2}{r_1^4}\right\} \quad (4.8)$$

The displacements are:

$$2G_1 u_x^{(1)} = \frac{K(K_1)-1}{2\pi((K_1+1)}\left\{\frac{(x-1)}{r_1^2} + A\big[(2-K_1)(x+a)-2a\big]\frac{1}{r_2^2} - 4A\frac{x(x+a)^2}{r_2^4}\right\} \quad (4.9)$$

$$2G_2 u_x^{(2)} = \frac{K(K_1-1)}{2\pi((K_1+1)}\left\{(A-1)\frac{1}{r_1^2}\right\} \quad (4.10)$$

$$2G_1 u_y^{(1)} = \frac{K(K_1-1)(y-b)}{2\pi((K_1+1)}\left\{\frac{1}{r_1^2} + \frac{AK_1}{r_2^2} - 4A\frac{x(x+a)}{r_4^2}\right\} \quad (4.11)$$

$$2G_2 u_y^{(2)} = \frac{K(K_1-1)(y-b)}{2\pi((K_1+1)}\left\{(1-A)\frac{1}{r_1^2}\right\} \quad (4.12)$$

The tractions at the interface are:

$$\left.\sigma_{xx}\right|_{x=0} = \frac{K(K_1-1)(1-A)}{2\pi(K_1+1)}\left\{\frac{1}{r_1^2} - \frac{2a^2}{r_1^4}\right\} \quad (4.13)$$

$$\left.\sigma_{xy}\right|_{x=0} = \frac{K(K_1-1)(y-b)}{\pi(K_1+1)}\left\{(1-A)\frac{a}{r_1^4}\right\} \quad (4.14)$$

4.2 Concentrated Movement

A Concentrated moment can be constructed by super-imposing a doublet with moment acting parallel to the interface onto another doublet with moment acting normal to the interface, such that all forces act in a counter-clockwise direction:

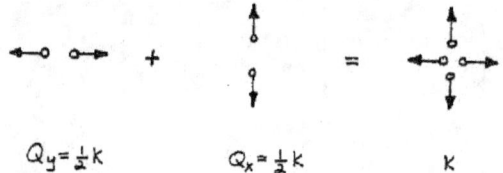

Figure 7

The Airy stress functions for the concentrated moment become the sums (3.4) with (3.21) and (3.5) with (3.22):

$$\mathbf{U}_1 = \frac{M}{2\pi}\left\{-\theta_1 - A\theta_2 A \sin 2\theta_2 + 2Aa\frac{\sin\theta_2}{r_2}\right\} \tag{4.15}$$

$$\mathbf{U}_2 = \frac{M}{2\pi}\left\{(A-1)\theta_1\right\} \tag{4.16}$$

The stresses are:

$$\sigma_{xy}^{(1)} = \frac{M}{2\pi}\left\{\frac{1}{r_1^2} - 2\frac{(x-a)^2}{r_1^4} - \frac{A}{r_2^2} + 2A[7(x+a) - 6a]\frac{x+a}{r_2^4} - 16A\frac{x(x+a)^3}{r_2^6}\right\} \tag{4.17}$$

$$\sigma_{xy}^{(2)} = \frac{M(A-1)}{2\pi}\left\{-\frac{1}{r_1^2} - 2\frac{(x-a)^2}{r_1^4}\right\} \tag{4.18}$$

$$\sigma_{xx}^{(1)} = \frac{M(y-b)}{2\pi}\left\{2\frac{(x-a)}{r_1^4} + 2A[-(x+a) + 2a]\frac{1}{r_2^4} + 16AA\frac{x(x+a)^2}{r_2^6}\right\} \tag{4.19}$$

$$\sigma_{xx}^{(2)} = \frac{M(y-b)(A-1)}{2\pi}\left\{-2\frac{(x-a)}{r_1^4}\right\} \tag{4.20}$$

$$\sigma_{yy}^{(1)} = \frac{M(y-b)}{2\pi}\left\{-2\frac{(x-a)}{r_1^4} + 2A[5(x+a) - 2a]\frac{1}{r_2^4} - 16AA\frac{x(x+a)^2}{r_2^6}\right\} \tag{4.21}$$

$$\sigma_{yy}^{(2)} = \frac{M(y-b)}{2\pi}\left\{2(A-1)\frac{(x-a)}{r_1^4}\right\} \tag{4.22}$$

The displacements are:

$$2G_1 u_x^{(1)} = \frac{M(y-b)}{2\pi}\left\{-\frac{1}{r_1^2} - \frac{AK_1}{r_2^2} - 4A\frac{x(x+a)}{r_2^4}\right\} \tag{4.23}$$

$$2G_2 u_x^{(2)} = \frac{M(y-b)}{2\pi}\left\{(A-1)\frac{1}{r_1^2}\right\} \tag{4.24}$$

$$2G_1 u_y^{(1)} = \frac{M}{2\pi}\left\{\frac{x-a}{r_1^2} + A\bigl[(-2-K_1)(x+a)+2a\bigr]\frac{1}{r_2^2} + 4A\frac{x(x+a)^2}{r_2^4}\right\} \tag{4.25}$$

$$2G_2 u_y^{(2)} = \frac{M}{2\pi}\left\{(1-A)\frac{x-a}{r_1^2}\right\} \tag{4.26}$$

The tractions at the interface are:

$$\left.\sigma_{xx}\right|_{x=0} = \frac{M(y-d)(A-1)}{\pi}\left\{\frac{a}{r_1^4}\right\} \tag{4.27}$$

$$\left.\sigma_{xy}\right|_{x=0} = \frac{M(A-1)}{2\pi}\left\{-\frac{1}{r_1^2} + 2\frac{a^2}{r_1^4}\right\} \tag{4.28}$$

4.3 Center of Shear Using Doublets with Moment

The center of shear shown below is constructed by superimposing a doublet with moment (counterclockwise) onto another doublet with moment (clockwise):

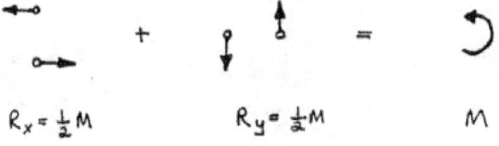

Figure 8

The Airy stress functions of the center of shear become the differences (3.21) from (3.4) and (3.22)from (3.5):

$$\mathbf{U}_1 = \frac{N}{2\pi(K_1+1)} \left\{ -\sin 2\theta_1 + (A-B)\theta_2 + A\sin 2\theta_2 - 2A\left[3\frac{\sin\theta_2}{r_2} + \frac{\sin 3\theta_2}{r_2}\right] - 2a\frac{\sin 2\theta_2}{r_2^2} \right\} \quad (4.29)$$

$$\mathbf{U}_2 = \frac{N}{2\pi(K_1+1)} \left\{ 2(B-A)a\frac{\sin\theta_1}{r_1} + (B-1)\sin 2\theta_1 + (B-A)\theta_1 \right\} \quad (4.30)$$

The stresses are:

$$\sigma_{xy}^{(1)} = \frac{N}{2\pi(K_1+1)} \left\{ -\frac{2}{r_1^2} + 16\frac{(x-a)^2}{r_1^6} + (A-B)\frac{1}{r_2^2} + 2\left[(-B-7a)(x+a)^2 + 84Aa(x+a) - 84Aa^2\right]\frac{1}{r_2^4} \right. \\ \left. + 16A\left[(x+a)^2 - 12a(x+a) + 12a^3\right]\frac{(x+a)^2}{r_2^6} - 192A\frac{ax(y-b)^4}{r_2^8} \right\} \quad (4.31)$$

$$\sigma_{xy}^{(2)} = \frac{N}{2\pi(K_1+1)} \left\{ -[(1-A)+(1-B)]\frac{1}{r_1^2} + 2\left[[(1-A)+7(1-B)](x-a) + 6(A-B)a\right]\frac{(x-a)}{r_1^4} \right. \\ \left. + 16A[(B-1)(x-a) + (B-A)a]\frac{(x-a)^3}{r_1^6} \right\} \quad (4.32)$$

$$\sigma_{xx}^{(1)} = \frac{N(y-b)}{\pi(K_1+1)} \left\{ -2\frac{(x-a)}{r_1^4} + 8\frac{(x-a)^3}{r_1^6} + [(B-A)(x+a) - 4Aa]\frac{1}{r_2^4} + A\left[-(x+a)^2 \right. \right. \\ \left. \left. + 8c(x+a) - 6a^2\right]\frac{(x+a)}{r_2^6} - 96A\frac{ac(x+a)(y-b)^2}{r_2^8} \right\} \quad (4.33)$$

$$\sigma_{xx}^{(2)} = \frac{N(y-b)}{\pi(K_1+1)} \left\{ -\left[[(1-A)+(1-B)](x-a)+2(A-B)a\right]\frac{1}{r_1^4} \right.$$
$$\left. + 8\left[(1-B)(x-a)+(A-B)a\right]\frac{(x-a)^2}{r_1^6} \right\} \tag{4.34}$$

$$\sigma_{yy}^{(1)} = \frac{N(y-b)}{\pi(K_1+1)} \left\{ 6\frac{(x-a)}{r_1^4} - 8\frac{(x-a)^3}{r_1^6} + \left[(-B-5A)(x+a)-4Aa\right]\frac{1}{r_2^4} + 8A\left[(x+a)^2 \right.\right.$$
$$\left.\left. - 4c(x+a)+6a^2\right]\frac{x+a}{r_2^6} + 96A\frac{ax(x+a)(y-b)^2}{r_2^8} \right\} \tag{4.35}$$

$$\sigma_{yy}^{(2)} = \frac{N(y-b)}{\pi(K_1+1)} \left\{ \left[[(1-A)+5(1-B)](x-a)+2(A-B)a\right]\frac{1}{r_1^4} \right.$$
$$\left. + 8\left[(B-1)(x-a)+(B-A)a\right]\frac{(x-a)^2}{r_1^6} \right\} \tag{4.36}$$

The displacements are:

$$2G_1 u_x^{(1)} = \frac{N(y-b)}{2\pi(K_1+1)} \left\{ (1-K_1)\frac{1}{r_1^2} - 4\frac{(x-a)^2}{r_1^4} + (AK_1-B)\frac{1}{r_2^2} \right.$$
$$\left. + 4A\left[(x+a)^2+(1-K_1)a(x+a)-2a^2\right]\frac{1}{r_2^4} - 32A\frac{ax(x+a)^2}{r_2^6} \right\} \tag{4.37}$$

$$2G_2 u_x^{(2)} = \frac{N(y-b)}{2\pi(K_1+1)} \left\{ \left[(1-A)+K_2(1-B)\right]\frac{1}{r_1^2} \right.$$
$$\left. + \left[(B-1)(x-a)+(B-A)a\right]\frac{(x-a)}{r_1^4} \right\} \tag{4.38}$$

$$2G_1 u_y^{(1)} = \frac{N}{2\pi(K_1+1)} \left\{ -(3-K_1)\frac{(x-a)}{r_1^2} + 4\frac{(x-a)^3}{r_1^4} + \left[(B+2A+AK_1)(x+a)+2A(K_1-1)a\right]\frac{1}{r_2^2} \right.$$
$$- 4A\left[(x+a)^2+(K_1+5)a(x+a)+6a^2\right]\frac{(x+a)}{r_2^4}$$
$$\left. + 32A\frac{ax(x+a)^3}{r_2^6} \right\} \tag{4.39}$$

$$2G_2 u_y^{(1)} = \frac{N}{2\pi(K_1+1)} \left\{ -\left[[(K_2+2)(1-B)+(1-A)](x-a)+2(A-B)a\right]\frac{1}{r_1^2} \right.$$
$$\left. + 4\left[(1-B)(x-a)+(A-B)a\right]\frac{(x-a)^2}{r_1^4} \right\} \tag{4.40}$$

The tractions at the interface are:

$$\sigma_{xx}\Big|_{x=0} = \frac{N(y-b)}{\pi(K_1+1)}\left\{[3(1-A)-2(1-B)]\frac{a}{r_1^4}+8(1-A)\frac{a^3}{r_1^6}\right\} \qquad (4.41)$$

$$\sigma_{xy}\Big|_{x=0} = \frac{N}{2\pi(K_1+1)}\Big\{-[(1-A)+(1-B)]\frac{1}{r_1^2}+2[(1-B)+7(1-A)]\frac{a^2}{r_1^4}$$
$$-16(1-A)\frac{a^4}{r_1^6}\Big\} \qquad (4.42)$$

4.4 Center of Shear Using Doublets without Moment

The center of shear shown below is constructed by superimposing a doublet without moment (forces acting radially outward) onto another doublet without moment (forces acting radially inward):

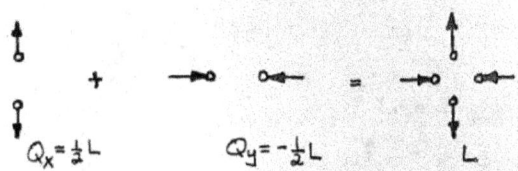

Figure 9

The Airy stress functions of the center of shear became the differences (2.22) from (2.4) and (2.23) from (2.5):

$$\mathbf{U}_1 = \frac{L}{2\pi(K_1+1)} \left\{ \cos 2\theta_1 + (B-A)\log r_2 + A\cos 2\theta_2 \right.$$
$$\left. - 2Aa\left(3\frac{\cos\theta_2}{r_2} + \frac{\cos 3\theta_2}{r_2} - 2a\frac{\cos 2\theta_2}{r_2^2}\right) \right\} \tag{4.43}$$

$$\mathbf{U}_2 = \frac{L}{2\pi(K_1+1)} \left\{ (B-A)\log r_1 + (1-B)\cos 2\theta_1 - 2(B-A)a\frac{\cos\theta_1}{r_1} \right\} \tag{4.44}$$

The stresses are:

$$\sigma_{xx}^{(1)} = \frac{L}{2\pi(K_1+1)} \left\{ 12\frac{(x-a)^2}{r_1^4} - 16\frac{(x-a)^4}{r_1^6} + (A-B)\frac{1}{r_2^2} + 2\left[(B+5A)(x+a)^2 - 12Aa^2\right]\frac{1}{r_2^4} \right.$$
$$\left. + 16A\left[-2(x+a) + 3a\right]\frac{(x+a)^3}{r_2^6} - 192\frac{ax(x+a)^2(y-b)^2}{r_2^8} \right\} \tag{4.45}$$

$$\sigma_{xx}^{(2)} = \frac{L}{2\pi(K_1+1)} \left\{ (A-B)\frac{1}{r_1^2} + 2\left[[(1-A)+5(1-B)](x-a) + 6(A-B)a\right]\frac{(x-a)}{r_1^4} \right.$$
$$\left. + 16\left[(B-1)(x-a) + (B-A)a\right]\frac{(x-a)^3}{r_1^6} \right\} \tag{4.46}$$

$$\sigma_{yy}^{(1)} = \frac{L}{2\pi(K_1+1)} \left\{ \frac{4}{r_1^2} - 20\frac{(x-a)^2}{r_1^6} + (3A+B)\frac{1}{r_2^2} - 2\left[(B+9A)(x+a)^2 + 24Aa(x+a) - 12Aa^2\right]\frac{1}{r_2^4} \right.$$
$$\left. + 16A\left[(x+a)^2 + 14a(x+a) - 12a^2\right]\frac{(x+a)^2}{r_2^6} - 192A\frac{ax(x+a)^4}{r_2^8} \right\} \tag{4.47}$$

26

$$\sigma_{yy}^{(2)} = \frac{L}{2\pi(K_1+1)} \left\{ [(1-A)+3(1-B)]\frac{1}{r_1^2} - 2\Big[[9(1-B)-(1-A)](x-a)+6(A-B)a\Big]\frac{x-a}{r_1^4} \right.$$
$$\left. + 16\big[(1-B)(x-a)-(B-A)a\big]\frac{(x-a)^3}{r_1^6} \right\} \tag{4.48}$$

$$\sigma_{xy}^{(1)} = \frac{L(y-b)}{\pi(K_1+1)} \left\{ 4\frac{(x-a)}{r_1^4} - 8\frac{(x-a)^3}{r_1^6} + (B+3a)\frac{x+a}{r_2^4} - 8A\big[(x+a)^2+6a(x+a)-6a^2\big]\frac{x+a}{r_2^6} \right.$$
$$\left. + 96A\frac{ax(x+a)^3}{r_2^8} \right\} \tag{4.49}$$

$$\sigma_{xy}^{(2)} = \frac{L(y-b)}{\pi(K_1+1)} \left\{ [(1-B)+(3-A)]\frac{x-a}{r_1^4} + 8\big[(B-1)(x-a)+(B-A)a\big]\frac{(x-a)^2}{r_1^6} \right\} \tag{4.50}$$

The displacements are:

$$2G_1 u_x^{(1)} = \frac{L}{2\pi(K_1+1)} \left\{ (K_1-3)\frac{(x-a)}{r_1^2} + 4\frac{(x-a)^3}{r_1^4} + \big[(AK_1-2A-B)(x+a)+2A(K_1+1)a\big]\frac{1}{r_2^2} \right.$$
$$+ 4A\big[(x+a)^2+(5-K_1)a(x+a)-6a^2\big]\frac{x+a}{r_2^4}$$
$$\left. - 32A\frac{ax(x+a)^3}{r_2^8} \right\} \tag{4.51}$$

$$2G_2 u_x^{(2)} = \frac{L}{2\pi(K_1+1)} \left\{ \Big[[(K_2-2)(1-B)-(1-A)](x-a)+2(B-A)a\Big]\frac{1}{r_1^2} \right.$$
$$\left. + 4\big[(1-B)(x-a)-(B-A)a\big]\frac{(x-a)^2}{r_1^4} \right\} \tag{4.52}$$

$$2G_1 u_y^{(1)} = \frac{L(y-b)}{2\pi(K_1+1)} \left\{ -(K_1+1)\frac{1}{r_1^2} + 4(x-a)^2\frac{1}{r_1^2} - (AK_1+B)\frac{1}{r_2^2} \right.$$
$$+ 4A\big[(x+a)^2+(K_1+1)a(x+a)-2a^2\big]\frac{1}{r_2^4}$$
$$\left. - 32A\frac{ax(x+a)^2}{r_2^6} \right\} \tag{4.53}$$

$$2G_2 u_y^{(2)} = \frac{L(y-b)}{2\pi(K_1+1)} \left\{ [(K_2+1)(B-1)+(A-B)]\frac{1}{r_1^2} + 4\big[(1-B)(x-a)+(A-B)a\big]\frac{x-a}{r_1^4} \right\} \tag{4.54}$$

The tractions at the interface are:

$$\sigma_{xx}\bigg|_{x=0} = \frac{L}{2\pi(K_1+1)}\left\{(A-B)\frac{1}{r_1^2} + 2[7(1-A)-(1-B)]\frac{a^2}{r_1^4} + 16(A-1)\frac{a^4}{r_1^6}\right\} \quad (4.55)$$

$$\sigma_{xy}\bigg|_{x=0} = \frac{aL(y-b)}{2\pi(K_1+1)}\left\{-[(1-B)+3(1-A)]\frac{1}{r_1^4} + 8(1-A)\frac{a^2}{r_1^6}\right\} \quad (4.56)$$

5 DISCUSSION OF RESULTS

All expressions for stresses, tractions at the interface, and displacements, were confirmed by checking the boundary conditions. For all the cases, at the boundary,

$$\sigma_{xy}^{(1)} = \sigma_{xy}^{(2)}$$
$$\sigma_{xx}^{(1)} = \sigma_{xx}^{(2)}$$
$$u_x^{(1)} = u_x^{(2)}$$
$$u_y^{(1)} = u_u^{(2)}$$

It was shown by Dundurs [6] that the normal stresses parallel to the interface must satisfy the following condition:

$$\sigma_{yy}^{(2)} = -\frac{2(\alpha - 2\beta)}{1-\alpha}\sigma_{xx}^{(1)} + \frac{1+\alpha}{1-\alpha}\sigma_{yy}^{(1)}$$

where

$$\alpha = \frac{\Gamma(K_1+1) - (K_2+1)}{\Gamma(K_1+1) + (K_2+1)}$$
$$\beta = \frac{\Gamma(K_1-1) - (K_2-1)}{\Gamma(K_1+1) + (K_2+1)}$$

This condition was also confirmed for all the cases.

It should be noted that Chapters 2 and 3 discuss four basic linearly independent elastic fields for doublets. Chapter 4 discusses superimposed doublets based on the four basic types. All the doublets presented in this work are first order doublets. The results given in this work can be used to derive the results for higher order doublets.

Appendices

A DISPLACEMENTS AND STRESSES FOR SOME BASIC AIRY STRESS FUNCTIONS[1]

χ	$2Gu_x$	$2Gu_y$	σ_{xx}	σ_{xy}	σ_{yy}
$r\log r\cos\theta$	$\frac{1}{2}(\kappa-1)\log r - \frac{x^2}{r^2}$	$\frac{1}{2}(\kappa+1)\theta - \frac{xy}{r^2}$	$\frac{x}{r^2} + \frac{2x^3}{r^4}$	$\frac{3x}{r^2} - \frac{2x^3}{r^4}$	$\frac{3x}{r^2} - \frac{2x^3}{r^4}$
$r\theta\sin\theta$	$\frac{1}{2}(\kappa+1)\log r - \frac{x^2}{r^2}$	$\frac{1}{2}(\kappa-1)\theta - \frac{xy}{r^2}$	$\frac{2x^3}{r^4}$	$\frac{2x^2y}{r^4}$	$\frac{2x}{r^2} - \frac{2x^3}{r^4}$
$\log r$	$-\frac{x}{r^2}$	$-\frac{y}{r^2}$	$-\frac{1}{r^2} + \frac{2x^2}{r^4}$	$\frac{2xy}{r^4}$	$\frac{1}{r^2} - \frac{2x^2}{r^4}$
$\cos 2\theta$	$-(3-\kappa)-\frac{x}{r^2}+\frac{4x^3}{r^4}$	$-(\kappa+1)\frac{y}{r^2}+\frac{4x^2y}{r^4}$	$\frac{12x^2}{r^4}-\frac{16x^4}{r^6}$	$\frac{8xy}{r^4}-\frac{16x^3y}{r^6}$	$\frac{4}{r^2}-\frac{20x^2}{r^4}+\frac{16x^4}{r^6}$
$\frac{\cos 2\theta}{r}$	$-\frac{1}{r^2}+\frac{2x^2}{r^4}$	$\frac{2xy}{r^4}$	$\frac{6x}{r^4}-\frac{8x^3}{r^6}$	$\frac{2y}{r^4}-\frac{8x^2y}{r^6}$	$-\frac{6x}{r^4}+\frac{8x^3}{r^6}$
$r\log r\sin\theta$	$\frac{1}{2}(\kappa+1)\theta-\frac{xy}{r^2}$	$\frac{1}{2}(\kappa-1)\log r + \frac{x^2}{r^2}$	$\frac{y}{r^2}+\frac{2x^2y}{r^4}$	$\frac{cx}{r^2}-\frac{2x^3}{r^4}$	$\frac{y}{r^2}-\frac{2x^2y}{r^4}$
$r\theta\cos\theta$	$\frac{1}{2}(\kappa-1)\theta+\frac{xy}{r^2}$	$-\frac{1}{2}(\kappa+1)\log r - \frac{x^2}{r^2}$	$-\frac{2x^2y}{r^4}$	$-\frac{2x}{r^2}+\frac{2x^3}{r^4}$	$-\frac{2y}{r^2}+\frac{2x^2y}{r^4}$
θ	$\frac{y}{r^2}$	$-\frac{x}{r^2}$	$\frac{2xy}{r^4}$	$-\frac{1}{r^2}+\frac{2x^2}{r^4}$	$\frac{2xy}{r^4}$
$\sin 2\theta$	$(\kappa-1)\frac{y}{r^2}+\frac{4x^2y}{r^4}$	$(3+\kappa)\frac{x}{r^2}-\frac{4x^3}{r^2}$	$\frac{4xy}{r^4}-\frac{16x^3y}{r^6}$	$\frac{2}{r^2}-\frac{16x^2}{r^4}+\frac{16x^4}{r^6}$	$-\frac{12xy}{r^4}+\frac{16x^3y}{r^6}$
$\frac{\sin\theta}{r}$	$\frac{2xy}{r^4}$	$\frac{1}{r^2}-\frac{2x^3}{r^4}$	$\frac{2y}{r^4}-\frac{8x^3}{r^6}$	$-\frac{6x}{r^4}+\frac{8x^3}{r^6}$	$-\frac{2y}{r^4}+\frac{8x^2y}{r^6}$
$\frac{\cos 3\theta}{r}$	$-(\kappa-2)\frac{1}{r^2}-2(8-\kappa)\frac{x^2}{r^4}$	$-2(\kappa+4)\frac{xy}{r^4}+16\frac{x^3y}{r^6}$	$-\frac{18x}{r^4}+\frac{8x^3}{r^6}+\frac{96x^3y^2}{r^8}$	$-\frac{6y}{r^4}-\frac{72x^2y}{r^6}+\frac{96x^4y}{r^8}$	$\frac{42x}{r^4}-\frac{136x^3}{r^6}+\frac{96x^5}{r^8}$
$\frac{\cos 2\theta}{r^2}$	$-\frac{6x}{r^4}+\frac{8x^3}{r^6}$	$\frac{2y}{r^4}-\frac{8x^2y}{r^6}$	$-\frac{6}{r^4}+\frac{48x^2}{r^6}-\frac{48x^2y^2}{r^8}$	$\frac{24xy}{r^6}-\frac{48x^3y}{r^8}$	$\frac{6}{r^4}-\frac{48x^2}{r^6}+\frac{48x^4}{r^8}$
$\frac{\sin 3\theta}{r}$	$2(\kappa-4)\frac{xy}{r^4}+16\frac{x^3y}{r^6}$	$-(2+\kappa)\frac{1}{r^2}+2(\kappa+8)\frac{x^2}{r^4}$	$-\frac{2y}{r^4}-\frac{40x^2y}{r^6}+\frac{96x^2y^3}{r^8}$	$\frac{66x}{r^4}-\frac{72x^3}{r^6}+\frac{96xy^4}{r^8}$	$\frac{24xy^3}{r^6}-\frac{48x^2y}{r^8}$
$\frac{\sin 2\theta}{r^2}$	$-\frac{2y}{r^4}+\frac{8x^2y}{r^6}$	$\frac{6x}{r^4}-\frac{8x^3}{r^6}$	$-\frac{24xy}{r^6}+\frac{48xy^3}{r^8}$	$\frac{42}{r^4}-\frac{48x^2}{r^6}+\frac{48y^2}{r^8}$	$\frac{10y}{r^4}+\frac{8x^2y}{r^6}-\frac{96x^2y^3}{r^8}$

Table 1

B DERIVATIVES OF TERMS IN THE AIRY STRESS FUNCTIONS

U	$U_{equivalent}$	$\frac{\partial U}{\partial a}$	$\frac{\partial U}{\partial b}$
$r_1\theta_1 \sin\theta_1$	$(y-b)\tan^{-1}\left(\frac{y-b}{x-a}\right)$	$-\frac{1}{2}\cos 2\theta_1$	$-\theta_1 - \frac{1}{2}\sin 2\theta_1$
$r_1 \log r_1 \cos\theta_1$	$(x-a)\log r_1$	$-\log r_1 - \frac{1}{2}\cos 2\theta_1$	$-\frac{1}{2}\sin 2\theta_1$
$r_2\theta_2 \sin\theta_2$	$(y-b)\tan^{-1}\left(\frac{y-b}{x+a}\right)$	$\frac{1}{2}\cos 2\theta_2$	$-\theta_2 - \frac{1}{2}\sin 2\theta_2$
$\log r_2$	$\log r_2$	$\frac{\cos\theta_2}{r_2}$	$-\frac{\sin\theta_2}{r_2}$
$r_2 \log r_2 \cos\theta_2$	$(x+a)\log r_2$	$\log r_2 + \frac{1}{2}\cos 2\theta_2$	$-\frac{1}{2}\sin 2\theta_2$
$\frac{\cos\theta_2}{r_2}$	$\frac{x+c}{r_2^2}$	$-\frac{\cos 2\theta_2}{r_2^2}$	$\frac{\sin\theta_1}{r_1}$
$\log r_1$	$\log r_1$	$-\frac{\cos\theta_1}{r_1}$	$-\frac{\sin\theta_1}{r_1}$
$\cos 2\theta_2$	$1 - 2\left(\frac{y-b}{r_2}\right)^2$	$\frac{\cos\theta_2}{r_2} - \frac{\cos 3\theta_2}{r_2}$	$\frac{\sin_t heta_2}{r_2} + \frac{\sin 3\theta_2}{r_2}$

Table 2

U	$U_{equivalent}$	$\frac{\partial U}{\partial a}$	$\frac{\partial U}{\partial b}$
$r_1\theta_1 \cos\theta_1$	$(x-a)\tan^{-1}\left(\frac{y-b}{x-a}\right)$	$-\theta - \frac{1}{2}\sin 2\theta_1$	$-\frac{1}{2}\cos 2\theta_1$
$r_1 \log r_1 \sin\theta_1$	$(y-b)\log r_1$	$-\frac{1}{2}\sin 2\theta_1$	$-\log r_1 - \frac{1}{2}\cos 2\theta_1$
$r_2\theta_2 \cos\theta_2$	$(x+a)\tan^{-1}\left(\frac{y-b}{x+a}\right)$	$\theta + 2 + \frac{1}{2}\sin 2\theta_2$	$-\frac{1}{2}\cos 2\theta_2$
$r_2 \log r_2 \sin\theta_2$	$(y-b)\log r_2$	$\frac{1}{2}\sin 2\theta_1$	$-\log r_2 + \frac{1}{2}\cos 2\theta_2$
θ_2	$\tan^{-1}\left(\frac{y-b}{x+a}\right)$	$\frac{\sin\theta_2}{r_2}$	$-\frac{\cos\theta}{r_2}$
$\sin 2\theta_2$	$2\frac{(x+a)(y-b)}{r_2^2}$	$\frac{\sin\theta_2}{r^2} - \frac{\sin 3\theta_2}{r_2}$	$-\frac{\cos\theta_2}{r^2} - \frac{\cos 3\theta_2}{r_2}$
$\frac{\sin\theta_2}{r_2}$	$\frac{(y-b)}{r_2^2}$	$-\frac{\sin 2\theta_2}{r_2^2}$	$-\frac{\cos 2\theta_2}{r_2^2}$
θ_1	$\tan^{-1}\left(\frac{y-b}{x-a}\right)$	$-\frac{\sin\theta_1}{r_1}$	$-\frac{\cos\theta_1}{r_1}$

Table 3

References

[1] J Dundurs and T Mura. Interaction between an edge dislocation and a circular inclusion. *Journal of the Mechanics and Physics of Solids*, 12(3):177–189, 1964.

[2] John Thomas Frasier. Force in the plane of two joined semi-infinite plates. *J. appl. Mech.*, 24:582–584, 1957.

[3] J Dundurs and M Hetenyi. The elastic plane with a circular insert, loaded by a radial force. *Journal of Applied Mechanics*, 28(1):103–111, 1961.

[4] M Hetenyi and J Dundurs. The elastic plane with a circular insert, loaded by a tangentially directed force. *Journal of Applied Mechanics*, 29(2):362–368, 1962.

[5] AEH Love. *The mathematical theory of elasticity*. 1927.

[6] J Dundurs. Discussion of a paper by d.b. bogy. *Journal of Applied Mechanics*, 36:650, 1969.

www.ingramcontent.com/pod-product-compliance
Lightning Source LLC
Chambersburg PA
CBHW081624220526
45468CB00010B/3012